Pierre Auguste Renoir

GREAT ART OF THE AGES

———————————

Pierre

GREAT ART OF THE AGES

Auguste Renoir

Text by MILTON S. FOX

Editor-in-Chief, The Library of Great Painters

Harry N. Abrams, Inc. Publishers, New York

ON THE COVER:
Luncheon of the Boating Party. 1881. Oil on canvas, 51 x 68"
Phillips Collection, Washington, D.C.

MILTON S. FOX, *Editor-in-Chief*
Standard Book Number: 8109-5133-9
Library of Congress Catalogue Card Number: 69-19714

Pierre Auguste Renoir

(1841–1919)

RENOIR'S ART IS THAT OF A MAN on good terms with life. His inspiration had its source in a sensuous delight in the world; and from his enchantment came a flow of paintings unrivaled in their lyricism. His art is never burdened with ideas or marked with conflict. As all true artists must, he transformed what he saw; but he saw with the eyes of a lover, and every element of his art affirmed his vision.

He was an amiable and uncomplicated man, free of disturbing tensions like those that give a sense of urgency, of painting as a refuge or comfort, to the works of some of his great contemporaries—Van Gogh, for example, or Degas or Toulouse-Lautrec. He painted with such relish that his

VENUS
Collection
Jacques Seligmann
& Co.

pleasure in his craft and in his subjects is at once communicated to us, and we are moved to say that no other artist has painted so freely, so spontaneously, and with so much joy. As a beginner, he was rebuked by Gleyre, a professor at the Ecole des Beaux Arts, with whom he had his only formal training in 1862 and 1863: "No doubt it's to amuse yourself that you are dabbling in paint?" And Renoir answered: "Why, certainly, and if it didn't amuse me to paint, I beg you to believe that I wouldn't do it."

There probably was no impudence in Renoir's answer: it must have struck him as strange that one might paint for any other reason. Indeed, as an old man he said: "I've had fun putting colors on canvas all my life"; and the two statements, like a pair of brackets, enclose a career that is so rare in balance and gratification that it does one good to contemplate it.

Renoir's art differs from that of other Impressionists: it is not simply a reflection of the pleasant aspects of life. We see it rather as the triumph of an ideal of happiness and beauty, and a victory over distress. It is remarkable that in the immense body of Renoir's work—some estimate it at around four thousand pic-

tures—there is no hint of darker, troubled moods. He had early years of hardship, made more disturbing by political attacks because he, as an Impressionist, did not conform to conservative standards of art; there was a period of self-doubt in the eighties and later—often disastrous for the artist; there were recurrent illnesses from the age of forty-one on, with the last fifteen years of his life a torment of arthritic pain. For a few years before his death he was so crippled that he had to be carried about when he was not in his wheel-chair, and his brush had to be strapped to his solidified hand. And yet, no trace of his personal suffering is anywhere to be seen in his work.

His art is always an affirmation. Instinctively he fixed on the happy aspects of the familiar things around him: the life of the streets, a beautiful day in the country; a bouquet of flowers or a profusion

OPPOSITE PAGE:
MADEMOISELLE ROMAINE LACAUX
Commentary on page 40

of fruits; people at play or making music; youth, and health. He was responsive to the delicate flower-freshness of small children, and he painted them as parents, with adoring incredulity, see their own. He had a remarkable flair for recording his time, and though his pictures never sink to the level of illustration, they still are a marvelous account of how people looked, what they wore, and what they did in their carefree moments. And the overwrought taste of the later nineteenth century in clothes and decor, has an agreeable look in Renoir's pictures, for he tempered its excesses while remaining true to its character.

Everything he touched he endowed with grace and charm. No other painter has discovered in the common actions of the human figure such vivacious gestures and attitudes, such piquant facial expressions. If Renoir is not one of the deepest of portrait painters—though he is one of the most engaging, showing us his sitters as a kindly host would see them in his parlor—he is unsurpassed as a painter of the human figure, clothed or unclothed. The female nude, above all, was his subject— he almost never painted the more angular, structurally obvious male nude—because in it he found delights of color and texture, forms which were ample, and warmth and fruitfulness and the promise of joy.

DANCE IN THE COUNTRY
Private Collection

Renoir started to earn money at painting when he was thirteen. His family had come to Paris from Limoges, where he was born, and Renoir was apprenticed as a painter of porcelain. Later he did pretty ornaments on fans and window shades, borrowing motifs from the French masters of the eighteenth century — Watteau, Boucher, Fragonard—whom he was to admire all his life. This early experience, no doubt, assured the decorative grace that is always present in his work, as well as that lustrous glow of his pigment which attracts us as color-music. The realistic painting of the sixties rooted him firmly in the substantial world; and as an Impressionist he discovered how to render the passing scene as spectacle, bathing his canvases in light, and filling them with a warmth of atmosphere that is all but palpable. He soon found that Impressionism was too limiting; subsequent works, reproduced here, show how he went beyond it. He never ceased to study the great art of the past, to which he was more closely attached than any artist of his generation—again an affirmation, this time of pleasure and nourishment found in the works of man. When asked where one

STUDY FOR
"THE BATHERS"
Private Collection
Courtesy M. Knoedler & Co.

learned to paint, he said, "In the museums, *parbleu!*" His final work—and here we mention also his beautiful sculpture—was in consonance with the spirit of twentieth-century painting; though he continued to image things, his lyricism was increasingly expansive and robust, inventive and daring, and Renoir came as close to an "abstract" coloristic art as his devotion to things he loved would permit.

His greatness springs from his color, of an unparalleled radiance and originality, which modeled the forms by subtle passage from nuance to nuance; a love of pigment as a magical substance; a wonderfully simplified drawing which especially in his more mature work banishes the angular, the harsh, and the rigid; and his instinct for decorative composition. Above all, there was his great love of people and the world; and the result was one of the most festive arts ever created.

PAINTED IN 1866

Spring Bouquet

OIL ON CANVAS, 41¾ x 32"

COURTESY OF THE FOGG ART MUSEUM, HARVARD UNIVERSITY

FOR ALL ITS APPARENT LOOSENESS, this painting has a precise structure. The flowers spill over into the lower left-hand corner in joyous profusion, with an asymmetry as free and wild as nature. But at once the artist proceeds to counter this unbalance. To the right of the vase, Renoir has developed a heavy shadow area, rich in purples, shaggy in contour, and sharply contrasted with the pool of light below it. Especially important is the placing of the ruler-straight lines in the lower right side of the canvas. In the set-up from which Renoir painted there was perhaps no such line as the diagonal which breaks into a zigzag in the extreme corner; yet something of the kind is necessary. If the reader will cover this diagonal he will see that the composition becomes lopsided.

This leads to an awareness of the basic structure. The composition is held in two triangles: a lower one with its apex at the rim of the bowl under the flowers; and the other with its apex at the top of the bouquet.

The Impressionist technique had not yet evolved when Renoir painted this picture. The brushing here is bold, the pigment fat and rich, suggesting the influence of Courbet, so important to the young painters of the sixties. And instead of a generalized effect of luminous color which will later *suggest* flowers rather than *depict* them, here the petals are separate and distinct.

Yet the canvas glows with light and color, indicating that Impressionism is just around the corner; there is something of that school in the feeling of the out-of-doors which Renoir has captured in this canvas. The texture of the flowers is marvelously rendered, and one is tempted to say that the perfume is there too.

The silvery-blue tonality, one of Renoir's most striking inventions in the work of this early period, is enlivened by accents of black, characteristically Renoir; the lovely sprinkling of yellow star-like blossoms is sheer delight. Renoir is at the threshold of his career, but his taste already is exceptional.

PAINTED IN 1867

Diana

OIL ON CANVAS, 75¾ x 50½″

NATIONAL GALLERY OF ART, WASHINGTON, D. C.
(Chester Dale Collection)

RENOIR TELLS that one day a man came to buy this canvas; they couldn't come to terms, for the prospective buyer wanted only the deer. The rest of the picture did not interest him!

The buxom lady became a Diana only through expediency. Renoir was a canny, realistic man as well as an artist. Here is what he said of this picture: "I intended to do nothing more than a study of a nude. But the picture was considered pretty improper, so I put a bow in the model's hand and a deer at her feet. I added the skin of an animal to make her nakedness seem less blatant—and the picture became a *Diana!*" He had his eye on the Salon of 1867; but the painting was rejected.

The canvas is one of the few on which Renoir employed the palette knife; it was at the time when the artist admired the thick impasto pigment and the prodigious realism of Courbet. The picture is clearly a studio piece: the lighting of the figure is harsh and the setting artificial.

Renoir is still trying to find himself, but his instinct and sensibility are sure. Again we have the cool silver-blue tones (one may recall that while Renoir paints cool at the start of his career and warm later, the reverse is the usual development). Throughout the painting there is a pronounced interest in varied textures, substances distinct in color, value, and rendering. Paint quality ranges from the smooth, unbroken sky to the enriched and dazzling flesh, then on to the more vigorously painted rocks, and finally to the broad, loose flecking of the foliage.

In a general way, the cool tones move from the upper left to the lower right corner, while the warmer tones are on the opposite axis. The bright band with which Renoir hoped "to make her nakedness less blatant" is part of the warm axis, but its lines follow the cool direction. There are many subtleties which the observer will discover: this is one of the great figure paintings of nineteenth-century realism, and its qualities are not easily exhausted.

PAINTED IN 1872

Pont Neuf

OIL ON CANVAS, 29¼ x 36½″

COLLECTION MRS. PAGE WRIGHT SMITH, PALM BEACH, FLORIDA

IN THIS MASTERPIECE of the painting of light, we have the happiest qualities of a bright summer day in Paris. How wonderfully Renoir has caught the vibrant many-sidedness of the city! In a most extraordinary way we are made to sense the large and simple compositional structure of the canvas, while at the same time we see the multitude of little details and contrasts of color and shape. The world is made of spots: buildings, windows, chimneys, flags, vehicles, statues, people; it is made of sky and clouds, light and air, of stone and water and foliage.

Despite the illusion of dazzling light, the picture is cool in tonality, even in the glare of the street. Yet within this larger simplicity, the whole palette is engaged: the yellows, reds, blues, violets, greens, blacks, and whites. There is an incredible richness of hue in every section: a good example is the varied blues of the bridge structure in the lower right.

The picture is divided into three big regions, each with its distinctive color, silhouette, and type of variation. The sky—a light zone—is spotted with irregular clouds, finely graded as to size, shape, and luminosity. The architectural mid-zone is the darkest: on the left a complicated grid of beautifully varied horizontal and vertical lines; to the right the broader elements of bridge and water, with strong diagonals leading to the extreme lower right corner. The lowest zone, the street, is the lightest of all, and here the promenaders and their shadows, scattered and free, repeat the verticals and diagonals of the mid-zone. If the reader will turn the picture upside down, these relationships may appear more clearly.

Renoir daringly makes the ground much whiter than the sun-drenched sky; he deliberately accepts the glare in his eyes to paint what is one of the first realistically back-lighted landscapes in art. He creates heat through coolness; the chaos of a busy street through order; and under his brush the grimy commonplaces of the city sparkle like jewels.

14

PAINTED IN 1874

The Loge

OIL ON CANVAS, 31 x 25″

COURTAULD INSTITUTE OF ART, LONDON

"BEAUTY," SAID STENDHAL, "is the promise of happiness." To Renoir, a simpler man, the words are synonymous: beauty is happiness. The promise and the realization are one. This picture is a hymn to the beauty of woman. It is an image of health, an exaltation of maturity, an idealization of togetherness. With what tact Renoir has placed the man in the background, covered half his face, and subdued the detail with which he is rendered! The woman is offered for full and rapturous gaze; her face and body and costume are more flower-like than the blossoms in her hair and corsage. She is a bouquet herself.

The man is Renoir's younger brother Edmond, who worshiped him. The woman is Nini Lopez, one of Renoir's models at the time, "with a profile of antique purity." This goddess chose to leave Renoir shortly after and marry a tenth-rate actor.

"Black is the queen of colors," said Renoir. A typically Renoir choice, it plays throughout, lustrous, patterned, varied, uniting the man and the woman. The alternate stripes stream down from the woman's bosom and face, like a radiation from the glamorous flesh.

The picture is full of Renoir's best qualities. Every aspect, by plan or by instinct, is harmoniously developed for visual delight. Take, for example, the pairing of things: the man and the woman; the gold of her bracelet and the opera glasses; the double gold stripe below her hand; the two pink blossom clusters on her bodice; the two vertical stripes in the light drapery; the twin pinks of her face and the flower above it; the warm spots of his face and gloved hand; even, as a final flourish, the two splashes of black on the ermine wrap next to his shirt front.

Again and again these paired attractions are related through short lower-right-to-upper-left diagonals; contrasting diagonals form a major movement upward and into the picture through the positions of the railing, the woman, and the man.

One's eyes keep wandering back to that lovely face, its doll-like perfection set off by playful wisps of hair which keep this beauty from cloying. If painting can be compared with music, surely this canvas is Mozartian.

16

PAINTED ABOUT 1876

Madame Henriot

OIL ON CANVAS, 27 x 21"

NATIONAL GALLERY OF ART, WASHINGTON, D. C.
(Gift of the Adele R. Levy Fund, Inc., 1961)

WHO ELSE BUT RENOIR could have given us an image so rare and charming in its delicacy and femininity? The picture is so like a happy dream that it is a little hard to think of it as a portrait of a real and mortal being. Other painters—Renoir's idols, Watteau, Boucher, and Fragonard, for instance—have given us charm and delicacy; but almost no one has succeeded so completely in etherealizing a personality, while yet keeping it close and warm and human.

We know that the subject is something really seen: this is Renoir's magic. The figure is painted with a minimum of substance, in colors atmospheric, diffuse, opalescent. With wonderful taste, Renoir gives fluidity to the lightness and bluishness of the whole, even though in order to do so he was obliged to sacrifice precision, as in the drawing of the hands and arms, the neck and shoulders.

In its high-keyed tonality the whole canvas seems a radiance, its luminosity reaching a climax in the head. Through the very dark accents of the golden brown hair and the sparkling brilliance of the eyes, Renoir further assures the dominance of the face. Through the lovely, shadowless porcelain colors of the skin, the suppression of the drawing of the nose except for the accents of the nostrils, the striking emphasis of the eyes, Renoir makes of the features a serene, decorative ensemble. Many of these devices have determined a mode of representation which has since been popularized by portraitists and fashion illustrators the world over. The superficialities can be copied; the poetic imagination which could conjure up such a vision is Renoir's alone.

The sitter, who often posed for Renoir at this time, was an actress of the Comédie Française; through Renoir's pictures of her she has become the image of womanhood at its loveliest.

PAINTED ABOUT 1879

The Umbrellas

OIL ON CANVAS, 71 x 45¼"

THE NATIONAL GALLERY, LONDON

SOMEONE HAS SAID that Renoir created a new mythology out of our poor humanity and endowed it with a sense of happiness. This picture, notable for its naturalness and exuberance, shows Renoir as the gentle, loving chronicler of everyday life. The busy commonplaces of existence captivate him, and under his brush the urban scene of his day is transfigured.

It is a showery spring day in Paris, and all the world is outdoors. Perhaps another sprinkling threatens, if one may judge by the half-hidden creature in the center of the picture and by the gallant at the left who is willing to share his umbrella with the girl in the foreground. But if anyone is dismayed by weather, he does not appear in this picture.

A silvery grey-lavender is the major color, and within its narrow range, has an extraordinary variety. Against this is played a dull gold quality, with sharper blues and greens and golden browns. And black. Always some spot of black, to make the ensemble sing.

Renoir delights in the curves of the umbrellas, and develops them throughout the picture: in the drawing of the girl in the foreground, in her bandbox, in the hoop, the children's bonnets, in the trees in the background.

The lovely *midinette* in front is surely one of the most winning figures ever painted. Drawn with beautiful simplicity, her lines reverberate in all parts of the composition.

In the right corner, Renoir has given us two doll-like youngsters of such charm as would justify a whole canvas devoted to themselves alone. They are united through their color, drawing, and the playful enrichments of their costumes, to the two women in blue. For this portion of the canvas, Renoir has reserved small delights for the eye and his richest color.

Renoir at this time was entering his "sour period," and here we can see what that term meant: an emphasis on edges, broader areas of color, a modeling like that in relief sculpture but with flattened faces; and distance achieved through line effects (as in the path into this picture). It is only the style which is tighter and drier; the joy of life, the transmutation into beauty of everything he touched—these are only the more in evidence.

20

PAINTED IN 1881

Luncheon of the Boating Party

OIL ON CANVAS, 51 x 68"

PHILLIPS COLLECTION, WASHINGTON, D. C.

NOT SINCE THE VENETIAN PAINTERS of the High Renaissance has the world seen such opulence in painting. But whereas the Venetians generally found their inspiration in the myths and lore of ancient times, Renoir's genius transmutes the common occurrences of everyday life into Olympian grandeur. These young gods and goddesses are friends of the painter, persons well known in Parisian art circles at the time. Aline Charigot, a favorite model whom Renoir married shortly after this picture was painted, sits at the left toying with the dog; the other girl at the table is another favorite model, Angèle, a lady of colorful repute; Caillebotte, wealthy engineer, talented spare time painter, who early began to acquire his great collection of Impressionist paintings which is now the pride of the Louvre—after a frenzy of opposition to the bequest in the nineties—Caillebotte sits astride the chair; the lady who so kittenishly closes her ears to a naughty jest is probably the actress, Jeanne Samary, painted by Renoir many times; the identity of most of the others is known.

One of the most notable features of the canvas is Renoir's felicity as an inventor of graceful, vivacious poses—poses which seem as though this is the way people ought to look. Delightful too is his knack for enlivening his canvases with piquant notes—a face emerging unexpectedly, a play of fingers, bits of still-life, bonnets, ribbons, beards, stripes, flowers. As is customary in Renoir's large compositions (and the Venetians') one side of the canvas is rich in things big and near; the other side presents a view into the distance, here a first-rate piece of Impressionist virtuosity.

Foreground and background are related in part by the awning, which in its striping combines the hues of the foliage with the warmer tones of the group; its playful serpentine edge echoes freely the curves in the group, and the breeze which flutters the valance sweeps also across the balcony. The feeling of animation is given in many subtle and striking ways: for example, the perspective of the balcony leads the eye to the upper right, but the open visual path into the distance offers an opposed attraction. And all the while the eye is cunningly led back, through the relationships of color—the spotting of blacks, for instance—and through line relationships, over backs, across heads, or following edges of color or light areas. And within every detail, no matter how small or casual, what a wonderful enrichment! This one canvas alone would be enough to assure a painter immortality.

PAINTED IN 1886

Mother and Child

OIL ON CANVAS, 32 x 25½″

COLLECTION CHESTER BEATTY, LONDON

IN THIS PICTURE—surely one of the sweetest presentations of the mother and child theme—there is no hint of the mawkishness which could have come from a more literal rendering. Everywhere, Renoir has simplified. Style has been imposed upon his vision—a style which transforms every line, every color, every circumstance of space, shape, or light.

We observe, to begin with, the contrasted techniques of the human figures and the surrounding landscape. The latter is an accomplished piece of Impressionist painting, its atmospheric tones bleached yet warm, its forms and textures of a caressing softness. In contrast, the mother and her child are drawn with the precise contours of Renoir's "sour period," and their surfaces are painted in more flat, unbroken tones. Yet their chalky coloring—also derived from the study of Italian fresco painting a few years earlier—and many shapes and lines, recall similar though weaker effects in the landscape; and one realizes how cunningly Renoir has brought into harmony the two antithetical manners.

As we study the figures, we observe what a brilliant draftsman was this artist who is more often celebrated for his color. The flattish areas of the white of the child's dress, the muted blue of the skirt, the tan of the jacket, and the flesh color—each has its distinctive contour and type of enrichment. Yet all are related: by edges which are continuous, by parallel directions, and by other correspondences—curves, points, and abrupt meanderings. Of particular interest are the intricate corkscrew silhouette of the child's legs and arm, and the related shapes of the mother's arm, breast, and head; and the way the breast and its nipple are almost exactly echoed in the drawing of the child's right thigh and hip, even to the point caused by the white skirt.

Renoir painted different versions of this picture of Mme. Renoir nursing their son Pierre, in 1885-86, and he took up the subject again after his wife's death in 1915, when he painted it in his loose, late style, and, with an assistant, sculptured it.

24

PAINTED IN 1884-87

The Bathers

OIL ON CANVAS, 45½ x 66″

PHILADELPHIA MUSEUM OF ART
(Bequest of Mrs. Carroll S. Tyson)

"AFTER THREE YEARS OF EXPERIMENTATION, *The Bathers,* which I considered my master-work, was finished. I sent it to an exhibition—and what a trouncing I got! This time, everybody, Huysmans in the forefront, agreed that I was really sunk; some even said that I was irresponsible. And God knows how I labored over it!"

The abuse which greets any new direction in art seems bizarre in retrospect; and though Renoir had often been attacked, he could hardly have expected derision for this masterpiece of his *période aigre* (sour period). It represents an amazing summation of everything he had done and learned. Here are the color and luminosity of a great Impressionist; drawing which results from his admiration of Ingres and Raphael; the benefits of his researches into the clarity and simplicity of fresco painting; the playful grace of his adored eighteenth-century French predecessors; and above all, that sweet ingenuousness which can exalt a bit of fun into something almost noble.

The charms of the picture are not confined to the ladies alone: few painters in the history of art could succeed like Renoir in matching the natural allurements of subject with the allurements provided by his own taste and style. The refined, melodic drawing—a violin-clarity of line—is one of the great achievements of art. It is especially important in this picture to savor the decorative silhouettes and spaces which Renoir has so brilliantly invented: the arabesque made by the contours of rocks and feet is an example. The intricacy of shape- and line-play is daringly counterpointed against the uncomplicated, cameo-smooth appearance of the figures themselves. Renoir here uses flat, unshadowed lighting, which ordinarily subdues modeling; and yet, through delicate tints, he produces a luscious roundness in the bodies.

The spirited poses were derived in part from the seventeenth-century bas-reliefs of Girardon, at Versailles; the piquant faces, the vivacious gestures, the robust elegance, are Renoir's. Cooks, housemaids, gamins, shopgirls—Renoir paints them, and we understand how it was that the gods of ancient times coveted mortal women.

PAINTED ABOUT 1890

In the Meadow

OIL ON CANVAS, 32 x 25¾″

THE METROPOLITAN MUSEUM OF ART, NEW YORK
(Bequest of Samuel A. Lewisohn, 1951)

AT THIS TIME Renoir was emerging from his *période aigre*. He returned temporarily to Impressionism, but it was an Impressionism wholly personal and original. With a new richness of color and a new vivacity in his brushwork, he paints thinly over light ground, creating a silken, undulant surface, like fine grass bending to summer breezes. The effect is one of peculiar radiance, as though beneath the thin but complex colors there was light and life. And while the figures are drawn with a fairly precise line—recalling his work of the eighties—we still remember pictures like this one for their color.

The painting has a simple basic structure—a kind of X, the two diagonals of which contrast in every respect. From the lower right to upper left a succession of pastel nuances moves from the hat in the corner, through the white dress and into the distant landscape and sky; this diagonal lies both on the surface of the design and in its depth. The other side of this X stays near the surface; it is made up of fuller colors, purples and corals. It follows the other girl's dress, from the lower left corner up into the tree, top right. Exquisite color contrasts are set up in all sections.

At the crossing of the X Renoir has placed a little bouquet of flowers in the hand of one of the girls. Again we are struck by the novelty—but complete naturalness—of the poses in Renoir's canvases: the girls have their backs to the observer, and they provide a human, but not personal, element.

Renoir's work in this period is sometimes criticized for "excessive softness of objects," as, for instance, in the foreground areas and in the trees. This would seem to be less a criticism than a misunderstanding. The artist was after precisely that effect of softness and fluidity, that melting quality that permeates the atmosphere on a warm summer day; the pleasant lassitude of such a day is caught here. In addition, Renoir was a constantly progressive artist, and this flowing color harmony, which here tends to be detached from descriptive functions, points to "abstract" color compositions in the early twentieth century.

PAINTED IN 1892

Two Girls at the Piano

OIL ON CANVAS, 45½ x 35½″

MUSEUM OF IMPRESSIONISM, THE LOUVRE, PARIS

IN THIS PICTURE we have a detailed and felicitous account of the surroundings of French family life at the end of the nineteenth century, and yet it does not break down into a clutter of things asking for attention. We see the piano with its candle holders, the chairs, the tasseled draperies, and in the room beyond we get a glimpse of a stuffy but inviting confusion. We may study the costumes and the hair-do which the girls wear, and on the piano is a typical "old fashioned" bouquet of the time.

This documentary character of Renoir's earlier work seldom conflicts with its qualities as art. The scene before us is bathed in a soft glow of light—its source and direction are not specific—and the rosy warmth helps us to join in the unselfconscious pleasure of the moment. This is a picture of leisure and relaxation and companionship. We feel this not only in the obvious indications of the subject, but more persuasively in the way the picture is put together.

The harsh, the rigid, and the angular are absent or suppressed; the gentle golden light which dilutes the colors brings them into the same family; and the ample forms which surround the girls seem to cushion and protect them. Yet Renoir here has given a surprising animation to a canvas which shows so little action. Our eyes are carried along sweeping lines and across forms which continue or stop one another. The foreground community of russet tones—the pillow on which the girl sits, the overstuffed chair with its music, and the piano—create an arc moving to our right and towards the wall; the bodies of the girls lean forward in a similar but opposed arc; and this arc is restated in the drapery. The relation of the three arms and hands in the center of the canvas is a masterly variation on this theme. The girls, a lovely range of delicate pastel tints, are a beautifully organized group, typically Renoir in their vivacious postures and facial expressions. Typically Renoir also is the brilliant clustering of the heads and ornamental details at a point where many lines converge from above and below.

PAINTED IN 1907

Gabrielle in an Open Blouse

OIL ON CANVAS, 25¾ x 21″

PRIVATE COLLECTION DURAND-RUEL, PARIS

POISE, YOUTH, AND EDEN-INNOCENCE: these Renoir gives us in a picture as bright and lovely as a bunch of freshly-picked garden flowers. The subject is probably Gabrielle, who is regarded with a special kind of affection by art-lovers everywhere. Shortly after the birth of the artist's second son, Jean (now a famous motion picture director), in 1893, Gabrielle was hired as his nurse—but not until she had met Renoir's usual basic condition for employment in his household: that she have a skin that "takes the light." How well she fulfilled this condition, the world knows.

In this picture she is not represented as a strongly individualized personality: what Renoir has painted can stand for everything sweet and artless in maidenhood which is not yet quite conscious of its womanliness. The head thrusts forward a bit stiffly, the bearing of the figure and the expression of the face suggests a girl who is hoping to be told that she sat very well—even though at the last moment she shied at complete nudity.

Renoir's taste is at its most exquisite here. Everything about the picture seems inevitable and right; there is nothing labored, nothing forced. The lively brushing of pearly greys and satiny whites gives the blouse the effect of a translucent cocoon, from which emerges the beautiful pink torso. The climax of color is in the face, broadly modeled and set off by the charming simplicity of the hairdress, the dark flow of which is in its own turn contrasted with the dainty blossom.

In the background there is nothing to distract from this delightful vision; instead, there is a field of turquoise blues, wonderfully varied in hue and texture. In a picture like this, we see a great artist in a relaxed mood, his inspiration pure, spontaneous, and free from any suggestion of artifice.

PAINTED ABOUT 1910

Woman at the Fountain

OIL ON CANVAS, 36 x 29″

PRIVATE COLLECTION, NEW YORK

THE GUIDING PRINCIPLE of most of Renoir's later work might be summed up in the word "breadth." He had already broadened the poses of his portrait figures until they occupied the whole of the canvas; now, in the 1900's, his color broadens out, and the small prismatic notes give way to simpler color areas in which a single hue is extended through all its nuances. The brush strokes themselves seem to have been made by a wider brush, and the stroke is less often a precise calligraphy: instead, there is something of a scrubbed effect.

The figures, as in the picture opposite, now are larger in their structure, and their curves are fuller and simpler: "correct" anatomy yields to amplitude. The figures suggest womanhood rather than femininity. The modeling, which in Renoir's earlier works often gave one the feeling that the other side of the figures might be flat, as in relief sculpture, now generously rounds the form. Even the conception of representation is broadened. For Renoir has painted this figure not as seen from a single, fixed viewpoint: the head is side view, the torso almost front view, one breast front view and the other side view, and the legs quite flat and also side view. In this connection, one is reminded of archaic art with its disregard for fixity of viewpoint.

The landscape, too, becomes more expansive, composed of fewer individual elements, and in keeping with this broadening out, Renoir's late figures and backgrounds flow into one another through blurred or ambiguous outlines. It is almost as though he is painting nature as a single, all-pervasive substance; and this substance gains articulateness through its more or less vague resemblance to trees or ground in some places, while elsewhere it resolves itself into the freely shaped forms of the human figure; but in essence it is the same basic stuff.

It seems logical that Renoir, finding the earth and its inhabitants beautiful, should arrive at a conception of the oneness of all nature.

34

PAINTED IN 1918

Girl with a Mandolin

OIL ON CANVAS, 25⅜ x 21¼″

PRIVATE COLLECTION DURAND-RUEL, PARIS

AT THE END OF HIS CAREER, famous and wealthy, recognized throughout the world as an artist of stature equal to that of the great painters of the past, Renoir was still painting with the joy and excitement of a young man who has just found his career. When a theme interested him he was content to paint it over and over again, for out of his own inner resources he could produce many original variations.

Several pictures of a girl playing a mandolin use similar props and resemble each other in the breadth of the drawing and in other ways. The present plate has, in general effect, Renoir's blonde, silver-gold tonality whereas others have more striking contrasts of color. This one seems freer in action and more serene, the lines and surfaces more flowing; in others we are aware of a greater density of forms, a heavier texture, and a certain compression.

Though the various elements may have been derived from the same corner of the studio, they are treated quite differently in each case. The wallpaper has a design of roses; in the present picture, Renoir has wittily set up an evenly spaced continuity between the patterned roses and the climactic real one in the girl's hair. The colors of the wall are also the colors of the diaphanous blouse; and the burnt almond of her skirt, used as an edging, helps to separate these color areas, while at the same time repeating the curved, linear motif. Against this overall similarity of hue and tonality Renoir singles out the deep pink flower, the yellow of the mandolin top, and the blue in the corner, for solo appearances. The dark notes in the very extreme right-hand corner, the mandolin, and the hair are sharply in contrast with the general filminess. As the eye goes from the dark spot in the corner to the darks of the mandolin and then to the hair, its motion is that of a long arc—and here again, we see how subtly Renoir relates elements of his picture.

PAINTED ABOUT 1884

Girl with a Straw Hat

OIL ON CANVAS, 21½ x 18½"

COLLECTION EDWIN C. VOGEL, NEW YORK

EARLY IN THE EIGHTIES Renoir became highly dissatisfied with the Impressionist method. He embarked on what has been called his "sour period"—the *période aigre*. This picture shows something of its qualities. We note the sharp drawing, the firm modeling. The face is smooth and plump and solid as a tight-skinned apple. The curves of the hat and dress are precise. The general effect of the picture reminds us of enamel. But Renoir is still the colorist, and it is from his subtle play of warm and cool tints—and not from shadow—that we get the feeling of roundness. The basic color scheme of the canvas is the opposition of complementaries, a burnished gold and a lapis-lazuli blue. There is a jewel-like quality here in the lustre of the yellows and in the depths of the blues. A brilliant variety of brush textures adds a further ornamental quality: the silken smooth hair which flows over the girl's shoulders, the fine lines like engraving in the treatment of the dress, the glittering effect in the hat, and the liquid sheen of the background.

ON PAGE 7:

PAINTED IN 1864

Mademoiselle Romaine Lacaux

OIL ON CANVAS, 32 x 25½"

THE CLEVELAND MUSEUM OF ART *(Gift of Hanna Fund)*

THIS SHINING PORTRAIT, painted at the age of only twenty-three, is perhaps the earliest signed and dated picture by Renoir which has come down to us. Like many a French artist before him, Renoir dwells on the ornamental grace of his subject. He is enchanted by the playful curves of the edging on the pinafore, and this becomes a theme which is varied in the contours of the hair, blouse, and skirt. Decorative enrichment of color led to the almost iridescent treatment of the background to the right; and again to the appealing invention of the hands resting on a cluster of flowers. Through passages of a creamy tint in the blouse, Renoir carries the eye upward to the head; the reds of the flowered background, the earrings and the lips lead us down again to the main statement of the red theme in the flowers. The canvas transmits in an amazing way the alert energy of the sitter, a delicate yet sure evocation of child-freshness.